PowerKids Readers:
Nature Books™

Vegetables

Jacqueline Dwyer

The Rosen Publishing Group's
PowerKids Press™
New York

1

Published in 2001 by The Rosen Publishing Group, Inc.
29 East 21st Street, New York, NY 10010

First Edition

Book Design: Michael de Guzman
Layout: Felicity Erwin, Nicholas Sciacca

Photo Credits: p. 1 © International Stock/Mike J. Howell; p. 5 © International Stock/J. Contreas Chacel; pp. 7, 9 © International Stock/Scott Campbell; pp. 11, 13, 15, 17 © Animals, Animals/Earth Scenes; p. 19 © International Stock/Nicole Katano; p. 21 © International Stock/Bob Schatz; p. 22 (dirt) © International Stock/Scott Barrow, (seeds) © Artville, LLC, (garden) © Animals, Animals/Earth Scenes/Mariann Slovak.

Dwyer, Jackie, 1970–
    Vegetables / by Jacqueline Dwyer.
      p. cm. — (PowerKids readers. Nature books)
    Summary: Describes a few different vegetables, how they grow, and why they are nutritious.
    Includes bibliographical references (p.).
    ISBN 0-8239-5679-2 (lib. bdg. : alk. paper)
    1. Vegetables—Juvenile literature. [1. Vegetables.] I. Title. II. Series.

SB324.D89 2000
635—dc21
                                                99-052529

Manufactured in the United States of America

Contents

A vegetable is a plant that we eat. Vegetables are many colors, shapes, and sizes.

A farmer plants seeds in the dirt. Most vegetables grow from tiny seeds. The little seeds grow into big vegetables.

There are lots of
vegetables. Potatoes,
peppers, and onions are
vegetables. Tomatoes are
vegetables, too.

Corn is a vegetable.
Vegetables grow in
different seasons.
Corn is a vegetable
that grows in the
spring and summer.

People use wooden stakes to help string bean plants grow tall. The plants climb up the stakes. They can grow up to six feet (1.8m) tall.

Sometimes small bugs eat the leaves of vegetable plants. A cucumber beetle is a small bug that eats the leaves of some vegetable plants.

15

Many people grow vegetables in gardens outside. They use special tools. A rake is a tool that breaks up the dirt. This makes it easier to plant seeds.

You can grow small vegetables at home. All vegetables need sunlight and water to grow.

It is good to eat vegetables every day. Vegetables make you grow strong. They taste great, too!

Words to Know

BUG

DIRT

FARMER

GARDEN

PLANT

RAKE

SEEDS

STAKE

VEGETABLES

Here are more books to read
about vegetables:
More Than Just a Vegetable Garden
by Dwight Kuhn
Silver Press

A Book of Vegetables
by Harriet L. Sobol
Dodd, Mead & Company

To learn more about vegetables, check out
this Web site:
http://www.kidsfood.org/kf_cyber.html

Index

Word Count: 180

Note to Librarians, Teachers, and Parents

PowerKids Readers (Nature Books) are specially designed to help emergent and beginning readers build their skills in reading for information. Simple vocabulary and concepts are paired with photographs of real kids in real-life situations or stunning, detailed images from the natural world around them. Readers will respond to written language by linking meaning with their own everyday experiences and observations. Sentences are short and simple, employing a basic vocabulary of sight words, as well as new words that describe objects or processes that take place in the natural world. Large type, clean design, and photographs corresponding directly to the text all help children to decipher meaning. Features such as a contents page, picture glossary, and index help children get the most out of PowerKids Readers. They also introduce children to the basic elements of a book, which they will encounter in their future reading experiences. Lists of related books and Web sites encourage kids to explore other sources and to continue the process of learning.